Report of Investigations 9683

Recommendations for Refuge Chamber Operations Training

Catherine Y. Kingsley Westerman, Ph.D., Kelly L. McNelis, Ph.D., and Katherine A. Margolis, Ph.D.

DEPARTMENT OF HEALTH AND HUMAN SERVICES
Centers for Disease Control and Prevention
National Institute for Occupational Safety and Health
Office of Mine Safety and Health Research
Pittsburgh, PA • Spokane, WA

June 2011

This document is in the public domain and may be freely copied or reprinted.

Disclaimer

Mention of any company or product does not constitute endorsement by the National Institute for Occupational Safety and Health (NIOSH). In addition, citations to Web sites external to NIOSH do not constitute NIOSH endorsement of the sponsoring organizations or their programs or products. Furthermore, NIOSH is not responsible for the content of these Web sites. The findings and conclusions in this report are those of the author(s) and do not necessarily represent the views of the National Institute for Occupational Safety and Health.

Ordering Information

To receive documents or other information about occupational safety and health topics, contact NIOSH at

> Telephone: **1–800–CDC–INFO** (1–800–232–4636)
> TTY: 1–888–232–6348
> e-mail: cdcinfo@cdc.gov
>
> or visit the NIOSH Web site at **www.cdc.gov/niosh**.

For a monthly update on news at NIOSH, subscribe to NIOSH *eNews* by visiting **www.cdc.gov/niosh/eNews**.

DHHS (NIOSH) Publication No. 2011–178

June 2011

SAFER • HEALTHIER • PEOPLE™

Contents

Abstract 1

Introduction 2

1. Gathering Information from Refuge Chamber Training Sessions 4

    List of Categories Used for Coding 5

    Results and Discussion 6

    Training Content 6

    Teaching Style 11

    Teaching Tools 11

2. Recommendations for Trainers 13

    Deployment 13

    Purge 14

    Oxygen 14

    Scrubber/Monitoring/Maintaining Air Quality 15

    Mine Communications System 15

    Location of Refuge Chamber in Mine 15

    Resources in Chamber 16

    Conserving Resources and Energy 16

    Troubleshooting 16

    The Unexpected 17

    Tips for Teaching: Advice for Inspiring Adult Learning 17

3. Information for Miners 22

    Top 20 Things to Know for Refuge Chamber Operation 22

Acknowledgments 23

References 24

Figures

Figure 1. Teaching techniques and activities. 12
Figure 2. Training model of an inflatable refuge chamber. 19
Figure 3. Mock-up of scrubber curtains and the stand used to hold them. 19
Figure 4. Mock-up of the internal oxygen controls and gauges on a refuge chamber. 20

Tables

Table 1. Mean and range for each category. (A higher mean represents a more detailed coverage of the given category.) 10

ACRONYMS AND ABBREVIATIONS

CO_2	carbon dioxide
DPOS	deploy purge oxygen scrubber
MSHA	Mine Safety and Health Administration
NIOSH	National Institute for Occupational Safety and Health
O_2	oxygen
SCSR	self-contained self-rescuer

Recommendations for Refuge Chamber Operations Training

Catherine Y. Kingsley Westerman, Ph.D.[1], Kelly L. McNelis, Ph.D.[1]
Katherine A. Margolis, Ph.D.[2]

Abstract

Refuge alternatives are airtight, reinforced shelters that underground coal miners can enter during a mine emergency. Although refuge alternatives are considered a last resort in an emergency, it is imperative that miners know how to operate these devices in the event of an emergency. This publication provides recommendations for training miners in how to operate a refuge chamber. It is organized into three sections. The first section details four refuge chamber training sessions that were observed by NIOSH researchers in order to determine the topics mine trainers were covering in their first efforts at refuge chamber training. The second section presents a guide for trainers which details the most important content to include in refuge chamber training and provides tips on teaching techniques and activities that may increase the effectiveness of the training. The third section presents a condensed version of the most important training content, specifically formatted for miners. Miners can refer to the list in the third section of this document to make sure they understand the basics of refuge chamber operation.

---

[1] Office of Mine Safety and Health Research, National Institute for Occupational Safety and Health.
[2] HealthEd Group, Inc. Formerly of OMSHR, NIOSH.

Introduction

Refuge alternatives are airtight, reinforced shelters that underground coal miners can enter during a mine emergency. Although different states and different mines refer to refuge alternatives by different names, this publication will refer to refuge alternatives that are close to the working face as refuge chambers, whether inflatable from a skid or constructed from steel. Other common terms for refuge chambers are rescue chambers, rescue shelters, and refuge shelters. Refuge alternatives that are outby of the face area, whether a prefabricated refuge chamber or one built into a crosscut, will be referred to as outby refuges because of their location in the mine. Outby refuges can be permanent, semipermanent, or portable and are usually located at every other self-contained self-rescuer (SCSR) cache. Outby refuges are sometimes called hardened rooms, outby shelters, or in-place shelters.

Refuge chambers are safe havens that provide breathable air, food, water, and a safe environment for up to 96 hours. They are typically made of steel or have tents that inflate from a steel skid. In 2008, the U.S. Department of Labor's Mine Safety and Health Administration (MSHA) mandated that all underground coal mines provide refuge alternatives at each working face and at additional locations outby the faces [73 Fed. Reg. 80698 (2008)].[*] Refuge chambers are usually portable so that they can be moved as mining advances.

It should be noted that entering a refuge chamber is a last resort for miners in an emergency situation. Although this option is considered a last resort, as refuge chambers are added to underground coal mines, mine trainers and refuge chamber manufacturers are faced with the task of training miners how to operate them. The regulation [73 Fed. Reg. 80698 (2008)] states that in addition to an introductory training session, each quarterly evacuation drill must include a review of the procedures for use of refuge alternatives. In addition, annual expectations training must include deployment and operation of refuge alternatives similar to those in use at the mine. As part of a larger project titled "Refuge Chamber Training,"[1] NIOSH researchers observed four introductory refuge chamber training sessions and created this document to summarize their findings and make recommendations for future training sessions.

This publication is intended to provide recommendations for training miners in how to operate a refuge chamber and may also be used to train miners on the operation of other types of refuge alternatives. It is organized into three sections. The first section details refuge chamber training sessions that were observed by NIOSH researchers in order to determine the topics mine trainers were covering in their first efforts at refuge chamber training. The second section presents a guide for trainers which details the most important content to include in refuge chamber training and provides tips on teaching techniques and activities that may increase the

---

[*]Federal Register. See Fed. Reg. in references.

[1]Additional refuge chamber training products resulting from this project include Harry's Hard Choices: Mine Refuge Chamber Training [NIOSH 2009a], Guidelines for Instructional Materials on Refuge Chamber Setup, Use, and Maintenance [NIOSH 2009b], Refuge Chamber Expectations Training [NIOSH 2009c], Emergency Escape and Refuge Alternatives [NIOSH 2010a], and How to Operate a Refuge Chamber: A Quick Start Guide [NIOSH 2010b].

effectiveness of the training. The third section presents a condensed version of the most important training content, specifically formatted for miners. Miners can refer to the list in the third section of this document to make sure they understand the basics of refuge chamber operation.

1. Gathering Information from Refuge Chamber Training Sessions

Four NIOSH researchers observed refuge chamber training sessions at four different underground coal mines. Because refuge chamber operations training is relatively new, observations were conducted to determine what mine trainers were covering in their first efforts at refuge chamber training. Mine emergency situations are unpredictable so it is crucial for miners to know how to operate their mine's refuge chambers. The existence of refuge chambers in the mine is only helpful to the extent that miners can operate them effectively if an emergency occurs, and the quality of training provided is one factor that will influence miners' ability to operate a refuge chamber. Observing current training sessions provided the opportunity to share mine trainers' teaching techniques and training materials with a wider audience and also to determine if there were any gaps in the training content where information should be included or further highlighted. These observations are discussed in this publication.

In order to quantify the training sessions, the research team selected 20 coding categories of information that may be included in refuge chamber training sessions based on a review of refuge chamber training materials and instructional videos and on a job task analysis. The job task analysis identified four basic steps for operating a refuge chamber that cut across the different types of refuge chambers; these include: (1) Deploy the unit, (2) Purge the refuge chamber, (3) activate the Oxygen, and (4) activate the Scrubber system, and are referred to collectively as DPOS or the Quick Start guide for refuge chamber operation. [NIOSH 2009b]. As the job task analysis was conducted, DPOS was conceived to simplify the process of operating a refuge chamber. However, it is also necessary for miners to know the details of operating a refuge chamber. The goal of this publication is to provide more detailed information on operating refuge chambers. To accomplish this goal, the 20 coding categories were comprised of the four steps of DPOS along with 16 other important steps in refuge chamber operation which were based on the review of refuge chamber training materials and instructional videos. Information from all of the coding categories is included in sections 2 and 3 of this document.

Of the 20 coding categories, 18 of these categories consisted of information necessary to explain the operation of refuge chambers. That is, the categories focused on the content required for a miner to learn how to operate the refuge chamber, perform any needed maintenance while inside the refuge chamber, and, in some cases, understand why certain steps should be taken when operating a refuge chamber. One of the 20 coding categories focused on presentation style of the training; specifically, this category focused on whether or not the trainer gave a preview of the training content. Another category focused on the physiological and psychological effects of being in a refuge chamber. Each of the 20 categories was condensed into concise phrases for the purpose of creating a brief document for use during coding of the training sessions. An initial draft of this document was pretested during a visit to a refuge chamber training session at a mine separate from the four observed in the main data collection. This pretest led to revisions to the categories themselves and refinement of the phrases and the scaling used to rate each category. The final list of phrases representing each category is shown below.

List of Categories Used for Coding

1. Deployment.
2. What to do with SCSR when entering chamber (keep it on until atmosphere is clear).
3. What to take inside/leave outside the refuge chamber (supplies, etc.).
4. Attaching/using mine communications devices with a refuge chamber.
5. Air lock/purging.
6. Explain concept of purge air (why is it important?).
7. Activating O_2 flow.
8. O_2 level (use chart for number of miners, location of chart).
9. Operating CO_2 scrubber (including adjusting for number of inhabitants).
10. How do you know the scrubber is working?
11. CO_2 (people generate CO_2 which can be an asphyxiant).
12. Heat and humidity, condensation inside chamber (they will develop).
13. Physiological / psychological issues.
14. Location of the resources in the chamber, specifically food, water, first-aid kit, etc.
15. Toilet: location, operation, where to put waste.
16. Conserve resources (cap lamps, body movement/energy, food/water).
17. Location of refuge chamber in mine.
18. Inspection and maintenance (any coverage of inspection information belongs here).
19. Preview of content/outcomes.
20. Why learning about refuge chamber operation is important.

Four NIOSH researchers were trained to code the refuge chamber training sessions using the twenty categories. The four coders practiced coding during three separate meetings prior to visiting the mines. These meetings involved coding a manufacturer-provided refuge chamber training video for the 20 categories followed by discussion of each coder's ratings and the reasoning behind each rating.

Following the meetings to practice coding, the coders visited actual refuge chamber training sessions at mines. They observed the training sessions and indicated the level of coverage of each category on a scale ranging from 1–5, starting at 1 and continuing toward 5 as the level of coverage of each category of information increased. Thus, if a category received no coverage, it would be rated a 1; if a category was mentioned but not explained or defined, it would be rated a 2; if a category was mentioned with a brief explanation or definition, it would be rated a 3; if a category was mentioned, defined, and/or steps associated with it were listed, it would be rated a 4; and finally, if a category was extensively covered, including additional descriptions such as a definition, steps needed to perform a task, explanation of the steps, or repetitive coverage, it would be rated a 5. Thus, lower numbers in this analysis indicate less coverage of a given topic and higher numbers indicate more extensive coverage. Because all of the categories were identified as important steps to refuge chamber operation, an ideal training session would cover each category extensively and receive a rating of 5.

Each coder coded independently during each refuge chamber training session. That is, the coders did not discuss their ratings with each other during the training session. The training content of each session was coded as it was mentioned by each trainer, which was not necessarily in the order listed on the coding sheet presented in the list above. In the section that follows each individual category and the ratings it received across all the observed training sessions will be discussed. Means and ranges for all 20 categories are presented in Table 1.

Results and Discussion

The NIOSH researchers observed 4 training sessions. Two of the training sessions were the initial training that miners received on refuge chambers and two of the groups had received prior training on this topic. Three of the training sessions were conducted by a mine trainer and one was conducted by a refuge chamber manufacturer. Although all four observed training sessions were for inflatable chambers, the material presented here applies to all types of refuge chambers. Three of these chambers contained curtain scrubbers and one contained a soda lime cartridge scrubber system. Both scrubber systems are designed to remove carbon dioxide from the air inside the chamber. Further explanation of CO_2 scrubbers will be provided later in this section.

Three training sessions were conducted for miners. During these training sessions, the average group included 16 miners with the group size ranging from 13 to 21. One training session was conducted with NIOSH researchers as the only audience members. The average length across all four training sessions was 71 minutes but times ranged from 30 to 97 minutes. Trainers presented the material in several ways: three trainers used computers, two provided handouts, two used a projection screen, three used PowerPoint presentations, two used a TV, and two used a DVD player. None of the trainers used posters.

Training Content

A detailed discussion of deployment is important during refuge chamber training because it is necessary to the operation of the refuge chamber. *Deployment is also the first step in the NIOSH Quick Start guide for operating a refuge chamber* [NIOSH 2010b]. For inflatable chambers, deployment includes pulling a lever to open the door to the tent, unrolling the tent, making sure the people are not in the space needed to inflate the tent, and filling the tent with air. For certain types of chambers, deployment also includes activating a strobe light on the exterior of the chamber. During the observed training sessions, trainers provided details and explanations on the deployment of refuge chambers. The average rating was 4.8, and scores ranged from 4–5. Trainers received a high rating because they were perceived as covering this category in extensive detail.

After miners deploy the refuge chamber they must decide what to take inside the chamber when they enter it and what to leave outside. This can include supplies, communications devices, tools, first-aid equipment (such as a stretcher), and other items. Covering topics in this training category increases the likelihood that miners will have what they need when inside the chamber and decreases the likelihood that they will bring something into the chamber that will hinder its operation. For inflatable chambers in particular, it is important to point out that sharp objects could puncture the tent if brought inside. Individual scores on this category ranged from 2–5, indicating that some of the coders felt that the trainers provided a lot of detail and explanation of what to bring inside/leave outside while other coders felt that the trainers provided minimal coverage. However, on average, coders rated trainers' discussions a 3.7, indicating a level of coverage higher than the midpoint of the scale for this category.

Trainers may also need to explain what to do with one's SCSR when entering a refuge chamber. Because refuge chambers are meant to be used during a mine emergency, it is probable that miners will be wearing their SCSRs when they approach and deploy the chamber. Miners should wear their SCSR into the chamber and keep it on until the atmosphere is clear as indicated by a gas monitor check. On average, trainers were rated a 2.2 on their coverage of this category; however, scores ranged from 1–5 indicating that there was not a consistent level of coverage at the different sessions.

In order to take full advantage of the refuge chamber, miners also need to know how to use mine communications devices while they are inside. This may include attaching existing communications devices to the chamber or operating communications devices provided inside the chamber. If miners know how to operate the communications devices, they will be able to communicate with rescue teams outside of the mine provided the communications system is still intact. Miners can then provide vital information, such as their location, an explanation of what occurred, and who is inside the chamber. Rescue teams can also relay information about the status of their rescue attempts. On average, trainers were rated 3.2 and scores ranged from 2–5 on coverage of this topic. All the trainers observed at least mentioned communications systems, but not all of them presented a detailed explanation of how to operate these systems. The importance of communications between underground and the surface in the case of an emergency cannot be stressed enough. Rescue efforts may be facilitated by information such as miners' location or gas monitor readings that can be communicated by miners in a refuge chamber. In future training sessions, trainers may want to cover this topic in more depth to increase the probability that miners will be able to communicate with the outside world during a mine emergency.

After miners deploy the refuge chamber they will need to know how to keep the breathable air inside the chamber from being contaminated by the air outside the chamber. This may include the use of an air lock or purge air system to maintain a safe atmosphere inside the chamber. *Activating the purge air is the second step in the NIOSH Quick Start guide for operating a refuge chamber* [NIOSH 2010b]. The exact system varies for different chambers, but regardless of the system, maintaining the clean air inside the chamber is important for successful operation of the refuge chamber. It is important to state how many miners the chamber can hold and how many miners can enter at a time. This information should be covered so miners are able to take full advantage of the refuge chamber and breathe safely while inside. On average, trainers were rated 3.8 on their coverage of how to operate the air lock or purge air system on their refuge chamber model. Scores ranged from 2–5; whereas some trainers covered this category in detail, others did not. It is important to cover this category in future training sessions.

Beyond explaining *how* to activate the air lock or purge air system, trainers can also explain *why* purge air is important. If miners understand that they are protecting the air inside the chamber from potentially dangerous gases outside the chamber, they are more likely to pay attention to the activation steps and remember them during an emergency. On average, trainers were rated 2.7 on coverage of this category, and scores ranged from 1–5. Therefore, some trainers were coded as not including an explanation of purge air and some trainers were coded as describing its purpose in detail. Covering the "why" behind the use of air locks and purge air systems is an important teaching tool to emphasize the importance of knowing how to operate these systems.

Also related to the air quality inside the refuge chamber, miners need to know how to activate the flow of oxygen. *This is the third step in the NIOSH Quick Start guide for operating a refuge chamber* [NIOSH 2010b]. Activating the oxygen is required to supply breathable air to the interior of the chamber. Scores ranged from 2–4 on this category; on average the coders rated the trainers 2.9. Thus, on average there was only a brief amount of coverage of this category in each of the training sessions attended. The level of coverage was relatively low for this important step in operating a refuge chamber.

In addition to knowing how to activate the oxygen flow, miners also need to be able to adjust the oxygen level based on the number of people inside the chamber. Most refuge chambers include a chart that details the amount of oxygen needed for a given number of occupants, and trainers were rated on their level of description of the chart and where to find the chart in the chamber. On average, trainers were rated 3.9 with scores ranging from 3–5. In general, trainers provided a reasonable amount of detail on this category.

People inhale oxygen and exhale carbon dioxide. Because carbon dioxide is an asphyxiant, refuge chambers were designed to remove carbon dioxide from the air inside the refuge chamber. In order to remove the carbon dioxide, refuge chambers include either curtain or pellet scrubbers. Miners need to know how to set up the scrubbers and adjust for the number of people inside the refuge chamber. *Operating the scrubber is the fourth step in the NIOSH Quick Start guide for operating a refuge chamber* [NIOSH 2010b]. Greater numbers of miners will generate greater amounts of carbon dioxide and the scrubbers need to be set up accordingly. Trainers were rated 4.2 on describing this process. Scores ranged from 3–5, indicating that coverage of this topic was detailed. Trainers can also teach miners how to tell if the scrubber is working properly. In some chambers, this is as simple as checking the color of the scrubber curtains (i.e., they will change color from blue to purple when spent). On average, the coders rated the trainers 2.7 on providing this explanation. Scores ranged from 1–5 indicating that some trainers were coded as not mentioning how to check the scrubbers at all and some trainers were coded as thoroughly describing how to perform this task. Finally, by explaining that carbon dioxide can be dangerous in high concentrations, trainers show why refuge chambers include a scrubber system, and why it is important to set it up correctly. Explaining that high concentrations of carbon dioxide are dangerous should provide motivation to miners to remember how to operate the scrubber. Trainers were rated 1.6 on their coverage of the importance of removing carbon dioxide from the air inside the chamber. Scores ranged from 1–2, indicating that overall coverage of this category was in the low range and not sufficient.

It is normal for heat, humidity, and condensation to occur inside the refuge chamber. If trainers explain this during refuge chamber training, miners will know that the chamber is operating correctly even though these conditions may occur when the miners are inside the chamber. On average, trainers were rated 1.7 in this category and scores ranged from 1–4. Some trainers did not mention heat, humidity, and condensation at all.

Miners can also expect physiological or psychological stresses to develop while inhabiting the refuge chamber. A mine emergency is a traumatic situation that may cause stress to develop. If trainers educate miners on what to expect in these situations and how to react to them, miners should be better prepared and more comfortable operating and living in the refuge chamber. None of the trainers mentioned physiological or psychological issues during the observed training sessions. All of the coders rated the trainers a 1 on this topic. Future training sessions should include detailed discussions on these very important topics. NIOSH has created a training program that specifically addresses the psychological and physiological issues related to refuge chamber use [NIOSH 2009c]. This program provides a valuable resource for dealing with these issues.

When miners are in the refuge chamber, it is important that they know the location of resources in the chamber, specifically the location of food, water, and the first-aid kit. The coders rated the trainers an average of 3.4 on discussing the location of these items. The ratings ranged from 2–5 indicating that some trainers were perceived as covering this in more detail than others. Trainers may also want to discuss the location and operation of the toilet in the chamber. Trainers were rated 3.4 on the discussion of this category, and scores ranged from 1–5, indicating that the coverage of this category was satisfactory but not excellent. Finally, it may also be important to discuss the conservation of resources in the chamber. Miners may have to remain inside the chamber for several days so they will need to ration the provided food and water as well as the light provided by their cap lamps. Miners may also want to limit the amount of energy that they spend moving around the chamber. Ratings on the discussion of these topics ranged from 1–4, and on average, trainers were rated 2.1 in covering this category. Overall, the coverage of this category was not particularly detailed.

Training sessions also provide an opportunity to discuss where refuge chambers are located in the mine and how to stay up-to-date on the location of portable chambers. If miners do not know where the refuge chambers are on any given day, they will likely not be able to use them in the event of an emergency. Thus, knowing where the refuge chambers are located and how to learn their location as they are moved is crucial to the miners in locating the refuge chambers during an emergency. During the observed training sessions, trainers were rated 3.3 on discussing this category, and scores ranged from 2–5. All trainers provided at least some mention of the location of the chamber and how to learn the location of the chamber; some of the trainers provided detailed information relating to this category.

Over time, the refuge chambers will need to be inspected and maintained. It is important to cover this information because without inspection and maintenance, the chambers may not work effectively when they are activated. Because of the unpredictability of emergencies, inspection and maintenance should be considered crucial topics. In addition, explaining these procedures to miners may give them a sense of respect for the chambers and thus discourage vandalism. Scores on this category ranged from 1–5, and on average, trainers were rated 3.9 on explaining these procedures. Some trainers did not mention these procedures at all, but others described them in great detail.

Table 1. Mean and range for each category. (A higher mean represents a more detailed coverage of the given category.)

Category	Mean	Range
Deployment	4.8	4–5
What to take inside/leave outside refuge chamber (supplies, etc.)	3.7	2–5
What to do with SCSR when entering chamber (keep it on until atmosphere is clear)	2.2	1–5
Attaching/using mine communications devices with refuge chamber	3.2	2–5
Air lock/purging	3.8	2–5
Explain concept of purge air (why is it important?)	2.7	1–5
Activating O_2 flow	2.9	2–4
O_2 level (Use chart for number of miners, location of chart)	3.9	3–5
Operating CO_2 scrubber (including adjusting for number of inhabitants)	4.2	3–5
How do you know the scrubber is working?	2.7	1–5
CO_2 (people generate CO_2 which can be an asphyxiant)	1.6	1–2
Heat and humidity, condensation inside chamber (they will develop)	1.7	1–4
Physiological/psychological issues	1.0	1
Location of the resources in the chamber, specifically food, water, first-aid kit, etc.	3.4	2–5
Toilet: location, operation, where to put waste	3.4	1–5
Conserve resources (cap lamps, body movement/energy, food/water)	2.1	1–4
Location of chamber in mine	3.3	2–5
Inspection and maintenance	3.9	1–5
Preview of content/outcomes	1.7	1–3
Why learning about refuge chamber operation is important	1.7	1–3

Teaching Style

In addition to the content of refuge chamber training, the way that trainers cover the training material is also an important factor in successful refuge chamber training. To quantify this aspect of training, two categories referring to teaching style were included in the coding categories. First, the coders indicated the degree to which the training session provided a preview of the content to be discussed and the outcomes which would result from the session. Scores ranged from 1–3, and the average rating was 1.7, which indicates that trainers did not provide much detail about what material would be covered and the outcomes of the session. Trainers should add a preview of the topics to be covered in future sessions and the expected outcomes so that miners know what to expect during the training session.

A second category related to teaching style is the inclusion of an explanation of why learning about refuge chambers is important. By explaining the importance of refuge chambers and the importance of learning how to operate them, trainers increase the likelihood that trainees will pay attention to the material and retain the information. On average, the trainers observed during these training sessions were rated a 1.7, and scores ranged 1–3. Overall, coverage of this category was not very detailed. Because the highest rating was only 3 on a 5-point scale, this may be an area for improvement. In the future, trainers may want to include a brief explanation or discussion of the importance of knowing how to operate a refuge chamber at the beginning of the training session. Descriptive statistics for these two teaching style categories are presented in Table 1.

Teaching Tools

The coders also made note of the teaching techniques and activities used by the trainers. A list of potential techniques and activities based in part on Lawson's [2009] active learning techniques was included in the coding sheet, and is presented in Figure 1. Rather than rating the level of usage of these tools, use of the tools was noted with a simple check box which was checked if the teaching tool was used and left blank if it was not used.

Some ways that trainers can involve miners in the training process are by providing a simulation, case study, field trip, story, or demonstration related to refuge chamber operation. A simulation is a description of a scenario involving refuge chambers where the participants must solve a problem of some sort. A case study is a detailed example of an event or situation where refuge chambers were used by actual people in a real emergency. A field trip is a trip to view a refuge chamber in place in a mine or in another related place such as on the grounds of the mine. Story-telling involves the use of real-life stories or personal experiences to illustrate points or develop rapport with the audience. A demonstration of how to use the refuge chamber could be performed on the refuge chamber itself at the mine or in a training facility or with a scaled-down model of part of the refuge chamber inside the classroom. During the observed training sessions, none of the trainers provided simulations, case studies, field trips, or stories. All four trainers provided demonstrations using either scale models of refuge chamber controls or a training unit. Examples of how to incorporate these techniques will be discussed in the section titled "Recommendations for Trainers."

Some activities designed to involve miners in the training session include discussion, small group activities, brainstorming, or use of a designated observer. Discussion is simply asking miners to participate by providing feedback, making comments, or asking questions. One of the four trainers included a discussion in the training session. Small group activities would involve breaking the miners up into smaller groups for discussion or another interactive activity such as role playing. Brainstorming is coming up with ideas for handling certain situations or about other issues. None of the trainers used small groups or brainstorming to reflect on the training material. Finally, a designated observer is a person who is specifically watching rather than doing; this person may be asked to perform some specific function such as observing whether an individual is doing a task properly. None of the trainers used designated observers to reflect on the training. These techniques could be added in the future to improve refuge chamber training. Examples of how to incorporate these techniques will be discussed in the section titled "Recommendations for Trainers."

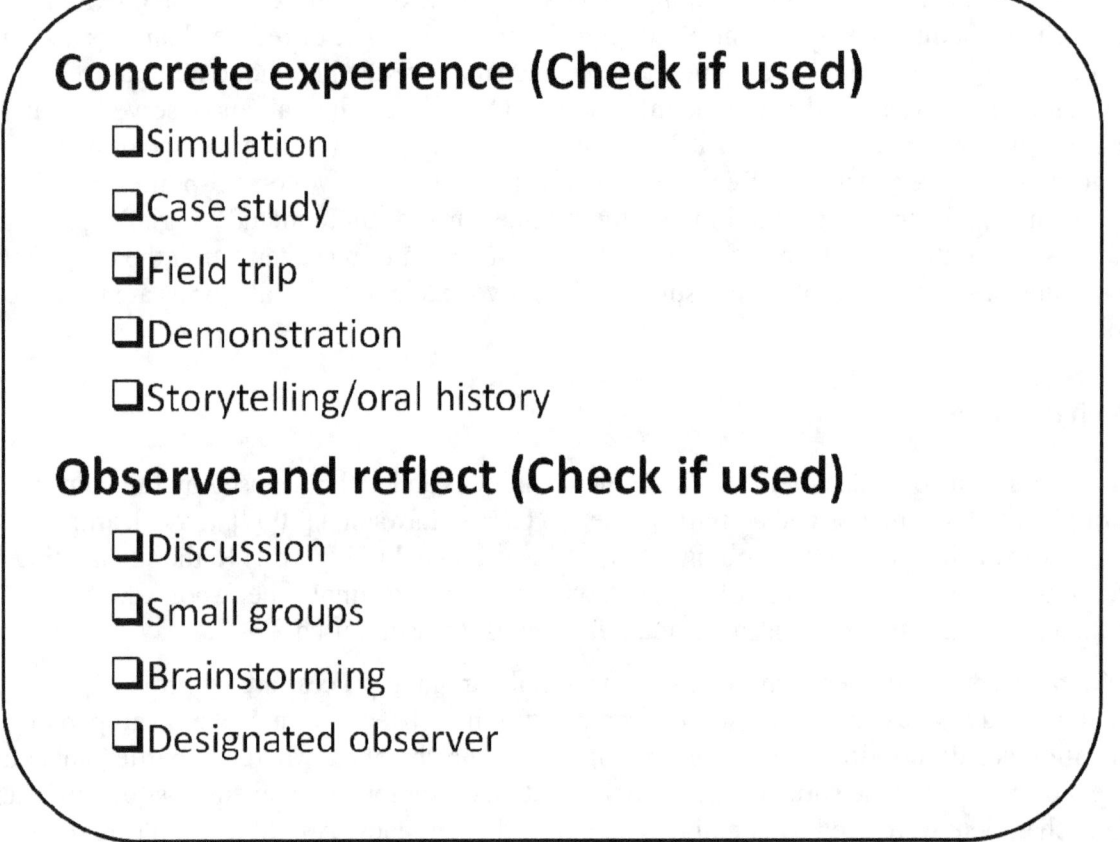

Figure 1. Teaching techniques and activities.

It is important to note that the observed mines may not represent all of the different types of refuge chamber training sessions that are being conducted in underground coal mines. Trainers are encouraged to consider the coding categories to evaluate refuge chamber training at their own mine.

2. Recommendations for Trainers

This section of the document is intended for trainers. It is not a free-standing training program; rather, this section will provide trainers with a starting point if they have not yet conducted any refuge chamber operations training or alternatively this section will highlight material to include in future training sessions. NIOSH researchers have created two PowerPoint documents [NIOSH 2010a; NIOSH 2010b] which provide a starting point for trainers who have not previously conducted refuge chamber training. *All the content listed below is crucial for miners to know in order to operate a refuge chamber.* Based on our observations, which are described in the first section of this document, some of this information may already be included in existing refuge chamber operations training. However, some of it may be missing from existing training. If one of these elements is not already in existing training, it should be incorporated. If it is only covered in part, or very briefly, trainers should ensure that coverage is increased or that all the elements included in this document are included in their training. In addition to content, we provide some recommendations for teaching strategies being used by mine trainers and also give examples of some strategies that are recommended for teaching adults new information. These strategies increase the likelihood that miners will retain information discussed during the training session.

The following section presents information that is crucial to refuge chamber operation. Each topic includes a short paragraph with relevant information for trainers followed by a bulleted list that provides a brief at-a-glance guide to the content.

Deployment

Deployment is the first step in the NIOSH Quick Start guide for refuge chamber operation [NIOSH 2010b]. If miners cannot deploy the chamber, they will not be able to use the chamber. Trainers should keep in mind that the steps to deployment may be different for each type of refuge chamber so they will need to tailor this material to their specific refuge chamber. Some of the essential elements to include are as follows:

- Methods of deployment.
- What to do with SCSR when deploying.
    - Continue wearing SCSR while deploying the chamber *and* inside the chamber until there is a safe, breathable atmosphere inside.
- What to take inside/leave outside chamber.
    - For example, leave sharp tools outside an inflatable chamber because of the danger of poking a hole in the tent.

Purge

Purging air is the second step in the NIOSH Quick Start guide for refuge chamber operation [NIOSH 2010b]. The specifics of purging the tent and/or the air lock are different for each type of refuge chamber. It may be helpful to explain the following:

- Methods of purging.
- Concept of purge air.
    o How purge air works and why it is necessary.
    o If using the chamber as a way station (a place to stop, rest, and regroup before moving on), the purge air may run out.
- Air lock/purging.
    o If miners are not careful when they are entering and leaving an air lock, they could contaminate the interior atmosphere of the chamber.

Oxygen

Activating the oxygen is the third step in the NIOSH Quick Start guide for refuge chamber operation [NIOSH 2010b]. Too much or too little oxygen can be dangerous, so it is important to get the proper flow. The following topics should be covered:

- Activating oxygen flow.
    o Without sufficient oxygen, the miners will not survive.
- Determining appropriate oxygen level for the number of miners in the chamber.

Scrubber/Monitoring/Maintaining Air Quality

Knowing how to operate the carbon dioxide (CO_2) scrubber is the fourth step in the NIOSH Quick Start guide for refuge chamber operation [NIOSH 2010b]. Just as it is important to know how to put oxygen into the chamber, it is also important to know how to remove the CO_2. Both are necessary for survival. The following topics should be covered:

- Steps to operating the scrubber.
- Explain why high CO_2 is dangerous.
    - CO_2 is dangerous in high concentrations.
    - CO_2 can kill if the scrubbers are not set up properly.
- Do not let scrubbers get soaked with water (e.g., if water pools on the floor, do not set scrubbers in the pool of water).
    - *Scrubbers will not work as well if they are soaked.*
- How do you know if it's working?
    - On some chambers, the scrubber curtains turn from blue to purple as they absorb CO_2.
    - In other chambers, miners may need to use a gas monitor to determine the level of CO_2 in the air.
- How and when to use the gas monitor in your chamber.

Mine Communications System

Establishing communication from underground to the surface is invaluable during an emergency situation. If the communications system is still working post-disaster, miners trapped underground can share information with the command center which may enable or speed their rescue. This could include information such as the miners' location or atmospheric and physical conditions in the mine. Because of this, it is important for miners to know:

- How to attach/use communications system (in some cases the mine communications system must be plugged into a port on the refuge chamber).

Location of Refuge Chamber in Mine

If miners do not know where a refuge chamber is located in the mine, they will not be able to use it. Therefore it is important to cover:

- Location of refuge chambers and outby refuges in the mine.
- How miners will know when the refuge chamber near the face is moved.
- How to find a refuge chamber in smoke or darkness.
    - Look for rigid 8-inch coil tactile signal installed on lifeline (this tactile signal indicates a refuge chamber is at the end of the branch line on which it is installed).

Resources in Chamber

Knowledge of resources in the chamber will be helpful because this will reduce confusion in an already-stressful situation. Trainers should discuss:

- Location of and how to operate resources
    - First aid
    - Toilet
    - Food
    - Water
    - Operations manual

Conserving Resources and Energy

Conserving resources and energy is another category of information that does not require very much time to cover but is important for miners to know if they have to enter a refuge chamber. Although they should conserve resources and energy, it is important that they move their bodies and muscles periodically to avoid medical complications that may result from being in the same position for long periods of time, such as muscle stiffness, body aches, and deep vein thrombosis [NIOSH 2009c].

- Conserve resources.
    - Cap lamps.
    - Food and water.
    - Supplies.
- Conserve energy but stretch and move around periodically.

Troubleshooting

A variety of troubleshooting topics are important to cover. Some are specific to a type of refuge chamber but they are all noted here regardless of type of chamber. If refuge chambers are not routinely maintained in good condition, they may not be ready or working when disaster strikes.

- Backup deployment strategy.
- Tent/tube leaks and how to fix them.
- Scrubber fan not working.
- Inspection and maintenance.

The Unexpected

The unexpected includes elements that miners may need to know but that may not be the most obvious things to explain to them. Knowing what to expect in advance could help miners remain calm when they are in the chamber. The following are examples of what to include:

- Condensation inside the chamber is normal.
- Heat and humidity may be present inside the chamber.
- Inflation noise (tent may screech while inflating).
- Zippers on tents may be hard to zip.

Tips for Teaching: Advice for Inspiring Adult Learning

Although the content of training is very important, the manner in which the training is presented is also an important factor in an effective training session. Among their principles of adult learning, Kowalski and Vaught [NIOSH 2002] state that training should be active and experience-based. Experiential learning is one technique that can be used in refuge chamber training to increase the likelihood that miners will retain the information that is covered. The key idea behind experiential learning is that "the most important learning comes from experience" [Rothwell 2008]. This idea fits well with refuge chamber operations training because most of the elements of operating a refuge chamber can be demonstrated and practiced. Experiential learning [Rothwell 2008] suggests that adults (in this case miners) may learn more easily by getting involved in the learning process through action-oriented activities and being given compelling reasons to learn. Also, receiving the material in several ways (e.g., verbally and visually) can help increase learning as stated by Laird's sensory stimulation theory [1985].

One teaching tool related to presentation style is the use of a preview at the beginning of the training session. This means giving the miners a list of 3–5 elements that will be included in the training. For example, a preview of refuge chamber training might include the following elements: (1) Introduction to refuge chambers, (2) Video on how to operate refuge chambers at the mine, (3) PowerPoint presentation reviewing steps for operating the mine's refuge chambers, (4) Hands-on demonstration in training unit, and (5) Questions and conclusion. Providing this information in advance helps the trainees follow along with the presentation and maintain their attention because they know what to expect. Second, explaining why it is important to take certain actions in operating a refuge chamber is another teaching tool discussed in the previous section. This is important because knowing why a certain step is important will motivate miners to perform the step. For example, explaining that carbon dioxide scrubbers will not work if they are sitting in puddles of water would likely provide motivation for miners to hang them up properly so they will work.

In addition to providing a brief preview and explaining why certain steps are important, it is also a good idea to address "What's in it for me?" This means showing the miners why it is important for each and every one of them to know how to use a refuge chamber and providing a compelling reason for them to learn as suggested by Rothwell [2008]. For example, the trainer could tell a story about a miner caught in an emergency by himself who could not rely on his fellow miners to help him. In this situation it would be important for the individual miner to know how to operate the refuge chamber. The trainer could also bring up the goal of getting everyone home safely to their families every day and explain why knowing how to use a refuge chamber could help achieve that goal. If miners understand why it is important for them to learn about refuge chamber operation, they may be more likely to pay attention to the content of the training and retain the information they learned.

Some experiential and active training techniques [Lawson 2009] that can be used effectively by mine trainers for refuge chamber operations training include: videos, role plays (includes designated observer), simulations, and storytelling. It is not necessary to use all these techniques at once, but combining them with lecture-style training will help break up the training and keep the attention of the miners. Here are some examples of how these different techniques could be used by a mine trainer:

First, some refuge chamber manufacturers provide videos which instruct trainees on how to operate their refuge chambers. This is a good option to use in a refuge chamber training session as a starting point or to reinforce a PowerPoint presentation which shows the steps involved in operating the refuge chamber. A number of the trainers observed used videos in combination with PowerPoint or verbal presentations, demonstrating that this technique should work with miners being trained.

Second, role playing could be used by having one miner perform some of the steps for operating a refuge chamber and having another miner provide feedback on whether or not the steps are being performed correctly. One of the mine trainers observed used "peer teaching" where two trainees paired up and one demonstrated the steps while the other critiqued and corrected the demonstrator. Afterwards, the two trainees switched roles. This training scenario enabled both trainees to go through two repetitions of the steps, and is essentially the same as a role play with one partner giving feedback. This is also an example of hands-on learning if mock-ups or training units can be used by the miners. All four of the observed trainers used some type of mock-up of the controls or training unit as part of their training. Figures 2–4 that follow show a training model of a refuge chamber, scrubber curtains and the stand on which they are to be hung, and a mock-up of the oxygen controls and gauges in a refuge chamber.

Figure 2. Training model of an inflatable refuge chamber.

Figure 3. Mock-up of scrubber curtains and the stand used to hold them.

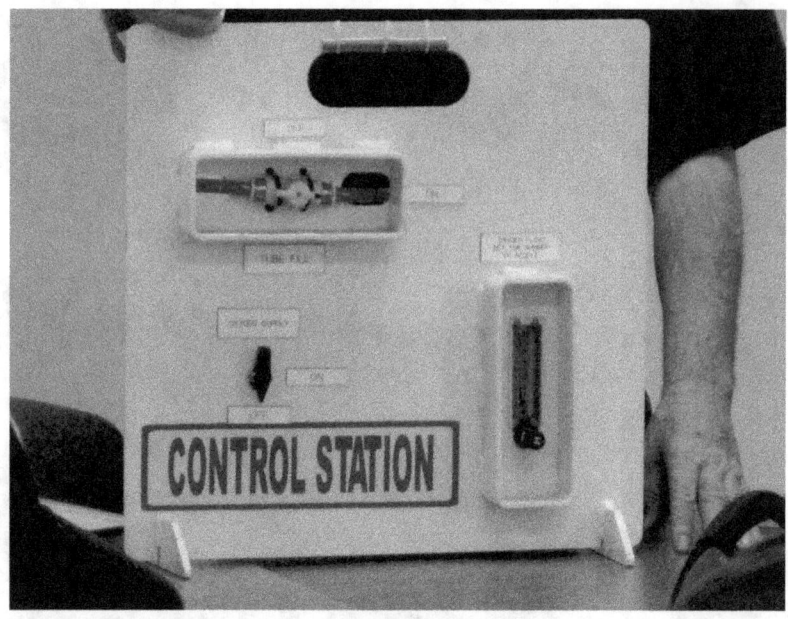

Figure 4. Mock-up of the internal oxygen controls and gauges on a refuge chamber.

A simulation is a more complex and involved form of role play. This could be used at the initial training for operating refuge chambers, or perhaps for annual refresher training when all the miners have already completed the initial training. For a simulation, a mine trainer could choose different groupings of miners in the training session and have them carry out a simulation of an emergency situation. The trainer might choose three miners and provide a scenario such as the following: "Imagine there was an explosion near your section of the mine and you three are the only survivors. You try to escape through all the different available routes out of the mine and discover that they are blocked. Now you have decided to go into a refuge chamber. One of you has a broken leg which you already splinted with some first-aid supplies, but the injured miner cannot move around very well. Work with your team to show what steps you need to take to get all three of you inside the chamber safely." This is an example of a small group activity and could also be a hands-on learning experience if a training unit or mock-ups of the controls are available.

The use of storytelling is an excellent way to reinforce the importance of knowing how to use a refuge chamber. If the trainer is an experienced miner or knows a miner who has been in an emergency situation, hearing a personal story from one of them may help convince miners of the importance of this information and encourage them to remember the steps for operating a refuge chamber. Use of a case study, which describes a real situation in detail, is another way to tell a story which shows miners what can happen to people just like them. Case studies of real accidents are available to the public in the Mine Safety and Health Administration's [MSHA 2010] digital library [http://www.msha.gov/TRAINING/LIBRARY/mshaPortal/index.html].

Activities such as those mentioned above are great ways to get miners involved in active and experiential learning. It may also be a good idea to have some kind of discussion of the different steps involved in operating a refuge chamber and why they are important. Trainers may tell the miners all the different steps multiple times, but in some cases it may enhance learning to get the miners talking about the steps they would take. This type of discussion could also involve brainstorming some ideas for what miners would do if they encountered certain problems, such as the unit not deploying properly. More specifically, brainstorming would work well with the troubleshooting content mentioned above.

Mine trainers should also use the available technology in their training sessions. For example, three of the training sessions observed used computers and PowerPoint presentations and two used handouts, TVs, and DVDs. It is not necessary to use all of these technologies, but if they are available, they may help maintain the attention of the miners, especially when used in conjunction with lecture-style training.

Finally, it is of the utmost importance for each mine trainer to know the details of the refuge chamber to be employed at his or her mine. This guide provides a general overview of the topics to be included in a refuge chamber training session, but it is up to the mine trainer to supply the details on the particular type of refuge chamber in their mine and other information such as the location of the chamber in the mine, how often it will be moved, and how miners will be notified of the new location when a refuge chamber is moved or a new one is installed.

3. Information for Miners

Top 20 Things to Know for Refuge Chamber Operation.

1. Know where refuge chambers are located.
2. Know how to find out when refuge chambers are moved.
3. Inspect, maintain, and respect the refuge chambers so they are ready for use at all times.
4. Deploy the chamber, and know the backup strategy for deployment.
5. Expect a screeching noise on deployment.
6. Keep your SCSR on until you check the chamber's atmosphere and determine it's safe.
7. Leave sharp tools outside; take first-aid kit inside.
8. Attach the mine communications system to the chamber.
9. Purge bad air from the air lock to keep the inside air clean.
10. Expect zippers on the air lock to be tight and hard to zip.
11. Activate oxygen and adjust level for the number of people.
12. Start the scrubber to remove carbon dioxide from the air.
13. Do not get the scrubber wet; it does not work when wet.
14. Check air to make sure the scrubber and oxygen are working properly.
15. Call outside on the mine phone if it is working.
16. Find the food, water, toilet, and first-aid kit inside the chamber.
17. Conserve cap lamps, food, water, and energy.
18. Change positions occasionally.
19. Know how to fix tent leaks and problems with the scrubber.
20. Expect heat, humidity, and condensation inside the chamber.

Acknowledgments

The authors would like to thank all of the miners and mine trainers who were observed for this project.

References

73 Fed. Reg 80698 [2008]. Mine Safety and Health Administration; refuge alternatives for underground coal mines; final rule (To be codified at 30 CFR Parts 7 and 75.).

Laird, D [2003]. Approaches to training and development. 3rd ed. Reading, MA: Addison-Wesley.

Lawson K [2009]. The trainer's handbook. 3rd ed. San Francisco, CA: Pfeiffer.

MSHA [2010]. Mine Health and Safety Administration Digital Library. [http://www.msha.gov/TRAINING/LIBRARY/mshaPortal/index.html]. Date accessed: November 1, 2010.

NIOSH [2009a]. Harry's hard choices: mine refuge chamber training. By Vaught C, Hall EE, Klein KA. Pittsburgh, PA: U.S. Department of Health and Human Services, Public Health Service, Centers for Disease Control and Prevention, National Institute for Occupational Safety and Health, DHHS (NIOSH) Publication No. 2009–122.

NIOSH [2009b]. Guidelines for instructional materials on refuge chamber setup, use, and maintenance. By Klein KA, Hall EE. Pittsburgh, PA: U.S. Department of Health and Human Services, Public Health Service, Centers for Disease Control and Prevention, National Institute for Occupational Safety and Health, DHHS (NIOSH) Publication No. 2009–148.

NIOSH [2009c]. Refuge chamber expectations training. By Margolis KA, Kowalski-Trakofler KM, Kingsley Westerman CY. Pittsburgh, PA: U.S. Department of Health and Human Services, Public Health Service, Centers for Disease Control and Prevention, National Institute for Occupational Safety and Health, DHHS (NIOSH) Publication No. 2010–100.

NIOSH [2010a]. Emergency escape and refuge alternatives. By Hall EE, Margolis KA. Pittsburgh, PA: U.S. Department of Health and Human Services, Public Health Service, Centers for Disease Control and Prevention, National Institute for Occupational Safety and Health, DHHS (NIOSH) Publication No. 2011–101.

NIOSH [2010b]. How to operate a refuge chamber: a quick start guide. By Hall EE, Margolis KA. Pittsburgh, PA: U.S. Department of Health and Human Services, Public Health Service, Centers for Disease Control and Prevention, National Institute for Occupational Safety and Health, DHHS (NIOSH) Publication No. 2011–100.

NIOSH [2002]. Principles of adult learning: application for mine trainers. By Kowalski KM, Vaught C. In: Peters RH, ed. Strategies for improving miners' training, Pittsburgh, PA: U.S. Department of Health and Human Services, Public Health Service, Centers for Disease Control and Prevention, National Institute for Occupational Safety and Health, DHHS (NIOSH) Publication No. 2002–156.

Rothwell WJ [2008]. Adult learning basics. Baltimore, MD: ASTD Press.

www.ingramcontent.com/pod-product-compliance
Lightning Source LLC
Chambersburg PA
CBHW081818170526
45167CB00008B/3450